***Checking the key features of a hoverfly*** – example: *Episyrphus balteatus* M…

## *Episyrphus balteatus*
Marmalade Hoverfly

W…
a… …nes are given.

**SIMILAR SPECIES** ***Here similar species are mentioned.***

***Hoverflies*** belong to the family Sylphidae in the flies Order of insects called Diptera which means two winged. Hoverfly species can range in size from 3mm to 20mm. Some specialise in imitating other insects such as bees and wasps. Many hoverfly larvae eat aphids, so encouging them into your garden by planting nectar-rich plants will benefit your garden by keeping aphid numbers down.

Like butterflies and moths the life-cycle of a hoverfly is egg, larva, pupa and adult

Female *Melanogaster hirtella* hoverfly egglaying

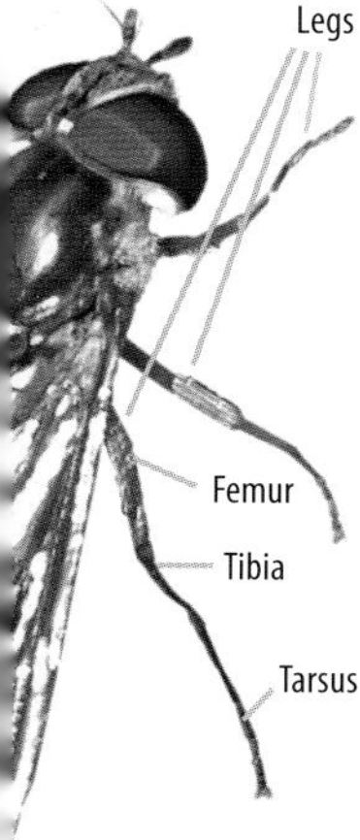


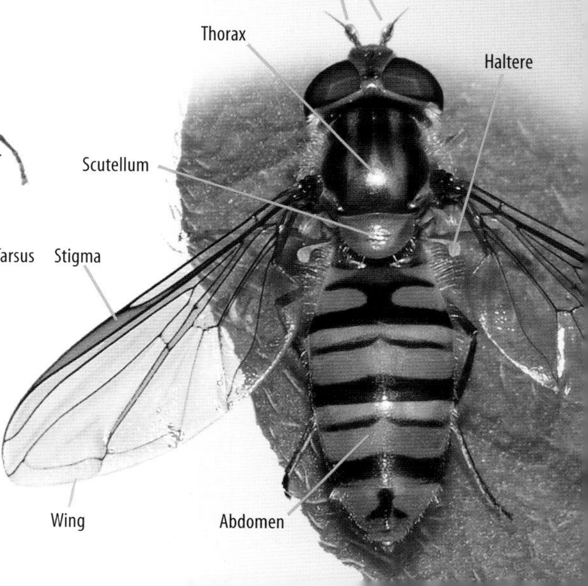

# Where to look

Hoverflies can be found in a wide variety of habitats: woodland, grassland, heathland, wetlands, gardens, verges and even waste ground. Some favour the same patch year after year, others travel further afield. The main habitats, and some of the best places to look locally, are mentioned here.

## Grassland and meadows

The chalk and limestone grasslands and the flower-rich meadows in our three counties are home to a variety of sun-loving hoverflies which are often seen visiting the many types of flowers found in these habitats. Some of the best sites to visit are – **Berks**: Seven Barrows and Watts Bank nature reserves (all BBOWT sites); **Bucks**: Dancersend, Bernwood Meadows, Pilch Field and the Upper Ray nature reserves (all BBOWT sites), Grangelands; **Oxon**: Sydlings Copse, Dry Sandford Pit, Chimney Meadows, Oxey Mead, Iffley Meadows, Hartslock, Chinnor Hill, Oakley Hill and Warburg nature reserves (all BBOWT sites), Aston Rowant National Nature Reserve and Watlington Hill. **LOOK FOR** *Paragus haemorrhous, Chrysotoxum bicinctum, Chrysotoxum cautum, Chrysotoxum festivum, Chrysotoxum verralli, Epistrophe grossulariae, Episyrphus balteatus, Scaeva pyrastri, Sphaerophoria scripta, Xanthogramma pedissequum, Xanthogramma citrofasciatum, Cheilosia bergenstammi, Cheilosia vulpina, Cheilosia urbana, Lejogaster metallina, Pipizella viduata, Pipiza austriaca, Volucella bombylans, Syritta pipiens*.

## Heathland and bog

Heathland and bogs, rare habitats in our three counties, are home to some specialist hoverfly species as well as more common species. Berkshire holds the vast majority of these special habitats in our area. **Berks**: Wildmoor Heath, Inkpen Common, Decoy Heath, Greenham Common and Snelsmore Common (all BBOWT sites). **LOOK FOR** *Paragus haemorrhous, Chrysotoxum festivum, Scaeva selenitica, Eristalis intricarius, Sericoyia silentis, Sericomyia lappona, Syritta pipiens*.

## Wetlands

Marshes and fens (especially the Cothill fens near Abingdon in Oxfordshire) are great places to visit to see some of the less often seen hoverflies– **Berks**: Hungerford Marsh, Thatcham Reedbeds; **Bucks**: Bernwood Meadows, Pilch Field and the Upper Ray nature reserves (all BBOWT sites); **Oxon**: Dry Sandford Pit, Parsonage Moor, Lashford Lane Fen, Sydlings Copse nature reserves (all BBOWT sites). **LOOK FOR** *Platycheirus clypeatus, Platycheirus granditarsus, Chrysotoxum verralli, Eupeodes latifasciatus, Cheilosia albipila, Melanogaster hirtella, Lejogaster metallina, Neoascia podagrica, Orthonevra nibilis, Eristalinus sepulchralis, Eristalis arbustorum, Eristalis intricarius, Helophilus pendulus, Helophilus trivittatus, Parhelophilus frutetorum, Syritta pipiens, Tropida scita*.



## Woodland

Our woodlands have a large number of attractive species of hoverfly, some of which are scarce in our three counties. They are often seen in woodland clearings or along woodland rides, either hovering in a sunlit patch, or visiting flowers (especially umbellifers such as hogweed or wild angelica). Woodland sites to visit are – **Berks**: Bowdown Woods and Moor Copse nature reserves (BBOWT sites), Windsor Forest; **Bucks**: Rushbeds Wood, Finemere Wood, Whitecross Green Wood, Little Linford Wood, Homefield Wood and Dancersend nature reserves (all BBOWT sites) and Bernwood Forest; **Oxon**: Foxholes, Sydlings Copse and Warburg nature reserves. **LOOK FOR** *Baccha elongata, Melanostoma scalare, Platycheirus albimanus, Chrysotoxum bicinctum, Chrysotoxum verralli, Dasysyrphus albostriatus, Dasysyrphus tricinctus, Epistrophe eligans, Epistrophe grossluariae, Epistrophe nitidicollis, Leucozona laternaria, Leucozona lucorum, Melangyna labiatarum, Meliscaeva cinctella, Scaeva pyrastri, Scaeva selenitica, Syrphus ribesii, Xanthogramma pedissequum, Callicera aurata, Cheilosia illustrata, Cheilosia pagana, Cheilosia fraterna, Cheilosia variabilis, Cheilosia antiqua, Ferdinandea cuprea, Rhingia campestris, Rhingia rostrata, Chrysogaster solstitialis, Neoascia podagrica, Orthonevra nobilis, Eristalis intricarius, Eristalis pertinax, Eristalis tenax, Myathropa florea, Parhelophilus frutetorum, Merodon equestris, Pipiza noctiluca, Pipiza austriaca, Sericomyia silentis, Volucella bombylans, Volucella inflata, Volucella pellucens, Volucella inanis, Volucella zonaria, Brachypalpoides lentus, Calyprobola speciosa, Criorhina floccosa, Criorhina ranunculi, Xylota segnis, Xylota sylvarum*.

## Gardens

Gardens can be a haven for many of our more common hoverflies. The larvae of many of these species will eat aphids, so it is worth encouraging hoverflies into your garden. One of the best ways to attract hoverflies to your patch is by planting nectar-rich flowers and shrubs. **LOOK FOR** *Melanostoma scalare, Platycheirus albimanus, Epistrophe eligans, Episyrphus balteatus, Eupeodes corallae, Eupeopdes luniger, Scaeva pyrastri, Sphaerophoria scripta, Syrphus ribesii, Rhingia campestris, Neoascia podagrica, Eristalis arbustorum, Eristalis tenax, Helophilus pendulus, Merodon equestris, Volucella inanis, Volucella zonaria, Syritta pipiens, Xylota segnis*.

4

# *Baccha elongata*

**DESCRIPTION** Wing length: 4–8mm. This small hoverfly is unmistakeable with its long and slender, wasp-waisted abdomen.

**WHERE TO LOOK** A common and widespread hoverfly of dappled woodland and hedgerows. Look out for this hoverfly as it flies low over vegetation. It visits flowers infrequently but often basks on sunlit leaves (wild garlic leaves and nettles are favourites). Head to any woodland with sunny rides in the three counties and, with patience, you should see this dainty little hoverfly. The following BBOWT natures reserves are all worth a visit – Bowdown Woods (Berks), Whitecross Green Wood and Rushbeds Wood (Bucks), Sydlings Copse and Warburg (Oxon).

**WHEN TO LOOK** April–November, peaking twice, in May and September.

**SIMILAR SPECIES**

***Sphegina clunipes*** (wing length: 5–7mm) and the tiny ***Neoascia podagrica*** (wing length: 3.5–5mm) (see p.43) do not have such long, narrow waists.



# *Melanostoma scalare*

**DESCRIPTION** Wing length: 5.5–8mm. This small hoverfly with a yellow-spotted abdomen is often one of the first hoverflies to be seen in the spring. The male (below) has a narrow abdomen with rectangular yellow spots, whereas the female's abdomen (left) is broader with triangular yellow markings.

**WHERE TO LOOK** A common and widespread hoverfly found throughout the three counties. It is found in woodland and scrub with lush vegetation, and will also come into gardens.

**WHEN TO LOOK** April–November, peaking twice, in May–early June and August.

**SIMILAR SPECIES**

***Melanostoma mellinum*** (wing length: 5–7mm) has a shorter abdomen and prefers grassland and heathland. Some yellow-spotted ***Platycheirus*** hoverflies look similar.

6

# *Platycheirus albimanus*

**DESCRIPTION** Wing length: 5–8mm. Most *Platycheirus* hoverflies have large yellowish abdominal spots, but *Platycheirus albimanus* is a grey spotted species. The spots on the male can appear bronze or yellowish at some angles, but in the female they always appear grey. The larvae feed on aphids.

**WHERE TO LOOK** Very common and widespread throughout the three counties. Found in woodland, scrub, along hedgerows and in gardens.

**WHEN TO LOOK** March–November, peaking twice, in May–early June and July–August.

**SIMILAR SPECIES**

The much less common ***Platycheirus ambiguus*** (wing length: 4.5–7mm), another grey-spotted species, prefers woodland margins and can be seen especially around blackthorn flowers.



# *Platycheirus clypeatus*

**DESCRIPTION** Wing length: 5–7.5mm. A small *Platycheirus* hoverfly with large yellowish abdominal spots. Like other species in the genus, they always settle with the wings folded over the abdomen so you need to look carefully through the wings to see the yellow spots.

**WHERE TO LOOK** A common and widespread hoverfly in Berks, Bucks and Oxon. It is found in both wet and dry grasslands and meadows, marshes and fens. It seems to hover continuously before settling briefly on rushes and sedges.

**WHEN TO LOOK** April–October, peaking in July–August.

**SIMILAR SPECIES**

***Melanostoma scalare*** (see p.5) and other yellow-spotted ***Platycheirus*** hoverflies.

# *Platycheirus granditarsus*

**DESCRIPTION** Wing length: 5–8.5mm. A large *Platycheirus* hoverfly with extensive orange markings on the abdomen. The legs are orange and black.

**WHERE TO LOOK** Widespread throughout Berks, Bucks and Oxon in marshy meadows, edges of ponds and ditches and other damp places. They often settle on rushes and sedges.

**WHEN TO LOOK** March–November, peaking twice, in May–early June and July–August.

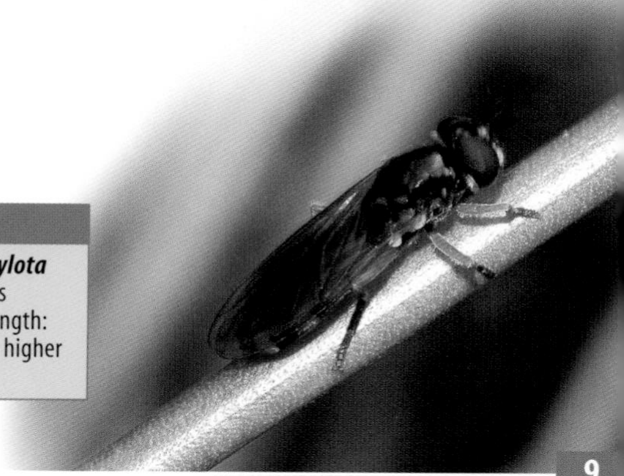

**SIMILAR SPECIES**

Could be confused with ***Xylota segnis*** (see p.71), but this hoverfly is larger (wing length: 7–9.5mm) and stands up higher on its legs.



# *Paragus haemorrhous*

**DESCRIPTION** Wing length: 3.5–4mm. A tiny black hoverfly, flying in sunny weather, often settling with the iridescent-looking wings open.

**WHERE TO LOOK** A widespread but local hoverfly in Berks, Bucks and Oxon, found in dry grasslands and heathland. Greenham Common in Berks is a good site for this species, especially by gorse bushes around which they constantly hover and occasionally settle. The limestone grassland at BBOWT's Dry Sandford Pit nature reserve (Oxon) is also worth checking out.

**WHEN TO LOOK** May–September, peaking in July.

**SIMILAR SPECIES**

The other ***Paragus*** hoverflies are very rare in Britain and have not been seen in our area.

10

## *Chrysotoxum bicinctum*

**DESCRIPTION** Wing length: 7–10mm. A black and yellow, sand wasp mimic. A distinctive hoverfly with long antennae, easily identified by the two bright yellow bands on the abdomen and the chocolate brown markings on the wings.

**WHERE TO LOOK** A fairly common hoverfly across Berks, Bucks and Oxon. Look for this hoverfly on umbellifer flowers in open grassland and along woodland rides. Bernwood Forest (Bucks) and BBOWT's Dry Sandford Pit nature reserve (Oxon) have an abundance of umbellifers where you are likely to see this species as well as a host of other hoverflies.

**WHEN TO LOOK** May–September, peaking mid-June–mid-July.

**SIMILAR SPECIES**

***Dasysyrphus tricinctus*** (see p.15), a woodland species, can look similar at certain angles but has one broad yellow band and two thinner bands.



## *Chrysotoxum cautum*

**DESCRIPTION** Wing length: 10–13mm. The largest *Chrysotoxum* species and a brilliant wasp mimic. The black and yellow striped abdomen and the slightly buzzing flight really does fool you and any potential predators into thinking that this insect is a wasp.

**WHERE TO LOOK** Widespread but not common in Berks, Bucks and Oxon. Look for this hoverfly in grassland and along hedgerows near to woodland. BBOWT's Sydlings Copse nature reserve and along the bridleway leading up to it is a good place to see this species.

**WHEN TO LOOK** May–September, peaking in June.

**SIMILAR SPECIES**

The rarer ***Chrysotoxum verralli*** (see p.13), is also a good wasp mimic but is smaller (wing length: 8–10mm).

# *Chrysotoxum festivum*

**DESCRIPTION** Wing length: 8–12mm. Another wasp mimic, this hoverfly has three narrow broken bars, curving towards the edges of the abdomen. Like other members of the genus it has long antennae.

**WHERE TO LOOK** Widespread but local across the three counties. Look for this hoverfly in grassy areas and heathland near to woodland edges. Bowdown Woods and Greenham Common (Berks) and Parsonage Moor (Oxon), all BBOWT sites, are worth checking out.

**WHEN TO LOOK** May–September, peaking in July and August.

**SIMILAR SPECIES**

Other ***Chrysotoxum*** species and the **Superb Ant-hill Hoverfly *Xanthogramma pedissequum*** (see p.31).



# *Chrysotoxum verralli*

**DESCRIPTION** Wing length: 8–10.5mm. Another wasp mimic with long antennae, wasp-like abdominal stripes and a buzzing flight.

**WHERE TO LOOK** A scarce species in the three counties. Found in grassland and woodland edges, sometimes in marshy areas. Locally, this species can be found flying around the scrubby margins of the fen at BBOWT'S Dry Sandford Pit nature reserve (Oxon).

**WHEN TO LOOK** June–October, peaking in July and August.

**SIMILAR SPECIES**

***Chrysotoxum cautum*** (see p.11) is larger (wing length: 10–13mm), but similarly marked.

## *Dasysyrphus albostriatus*

**DESCRIPTION** Wing length: 6–9.5mm. A wasp mimic with slanting yellow bars on the abdomen and two grey bars on the thorax.
**WHERE TO LOOK** Widespread but local in the three counties. Found along woodland edges, often on umbellifers. Look out for this species basking on sunlit vegetation especially on hot spring days at Bowdown Woods (Berks) and at Dry Sandford Pit (Oxon), both BBOWT nature reserves.
**WHEN TO LOOK** April–November, peaking twice, in late May–June and late August–early September.

**SIMILAR SPECIES**

***Dasysyrphus venustus*** (wing length: 9.5–11.5mm), another woodland species, is larger, but has slightly hooked bars on the abdomen and no grey bars on the thorax.

## *Dasysyrphus tricinctus*

**DESCRIPTION** Wing length: 7–10mm. This wasp mimic has distinctive triple bars on the abdomen, the first bar broad, the other two thin.
**WHERE TO LOOK** A hoverfly mainly seen in spring along woodland edges, though never in any great numbers. Another species to look out for basking on sunlit vegetation at BBOWT's Bowdown Woods nature reserve in Berks.
**WHEN TO LOOK** March–October, peaking twice, in late May–early June and late August–early September.

**SIMILAR SPECIES**

***Chrysotoxum bicinctum*** (see p.11) has two broad yellow bars on the abdomen and chocolate markings on the wings.

# *Epistrophe eligans*

**DESCRIPTION** Wing length: 6–9.5mm. This small hoverfly is seen almost constantly hovering by low leafy branches. When it does eventually settle on a leaf its likeness to a small honey bee becomes apparent. The abdomen has two triangular markings and a number of yellowish bands which can vary from none at all to three. The legs are entirely yellow.

**WHERE TO LOOK** A common hoverfly of sunny woodland edges, scrub and garden with trees and shrubs, it is especially fond of hawthorn flowers. Look out for this species in spring in woodlands anywhere in Berks, Bucks and Oxon.

**WHEN TO LOOK** March–August, peaking in May.

**SIMILAR SPECIES**

***Eristalis*** species (see pp.45–49) can have similar markings but are much larger.



# *Epistrophe grossulariae*

**DESCRIPTION** Wing length: 9–12mm. A medium-sized wasp mimic hoverfly with distinctive parallel-sided yellow bands on the abdomen and dully greyish stripes on the thorax.

**WHERE TO LOOK** Widespread but local in Berks, Bucks and Oxon. A hoverfly of woodland edges, look for this species on umbellifer flowers and, in nearby meadows and grasslands, on knapweeds. Try BBOWT's Sydlings Copse or Parsonage Moor nature reserves in Oxon, but it could turn up anywhere if the habitat requirements are right.

**WHEN TO LOOK** May–October, peaking in July.

**SIMILAR SPECIES**

***Epistrophe nitidicollis*** (wing length: 8–11mm), found in woodland rides and glades, looks similar but has orange antennae.

# ***Episyrphus balteatus*** Marmalade Hoverfly

**DESCRIPTION** Wing length: 9–12mm. One of our most common and well-known hoverflies. The double black bands on each abdominal segment are unique, though the abdomen colour can range from the typical orange colour to darker brown.

**WHERE TO LOOK** This familiar hoverfly is the most frequently recorded species in Britain. It is often abundant on flowers in gardens, meadows, woodland rides and other habitats throughout Berks, Bucks and Oxon.

**WHEN TO LOOK** All year, peaking in July and August.

**DID YOU KNOW?**

This hoverfly gets its common name, Marmalade Hoverfly, from its typical orange marmalade colour. Plant some nectar-rich flowers to encourage this hoverfly into your garden, their larvae are voracious aphid eaters.



# ***Eupeodes corallae***

**DESCRIPTION** Wing length: 5–8mm. A small hoverfly with an oval black and yellow abdomen. Very variable, the yellow to orange spots are usually separated in the centre but can be fused together.

**WHERE TO LOOK** Another common and widespread hoverfly in the three counties whose numbers each year are boosted by migrants. It can be found in a variety of open habitats, including gardens.

**WHEN TO LOOK** March–November, peaking in July and August.

**SIMILAR SPECIES**

***Eupeodes latifasciatus*** (see p.21) is slightly larger (wing length: 6.5–8.5mm) and has yellow markings on the abdomen more often fused together, especially in the males.

## *Eupeodes latifasciatus*

**DESCRIPTION** Wing length: 6.5–8.5mm. A small hoverfly with an oval black and yellow abdomen. Very variable, the yellow to orange spots are usually separated in the centre but can be fused together.

**WHERE TO LOOK** Widespread in the three counties, its numbers can fluctuate dramatically from year to year. It can be found in a variety of open habitats, with a preference for wetter sites with rushes and lush vegetation. The fenland reserves near Abingdon in Oxon – the Ruskin Reserve and BBOWT's Dry Sandford Pit, Parsonage Moor and Lashford Lane Fen – are ideal places to look, as is Greenham Common in Berks.

**WHEN TO LOOK** March–October, peaking in August.

**SIMILAR SPECIES**

***Eupeodes corallae*** (see p.19) has yellow markings more often separated in the centre of the abdomen.



## *Eupeodes luniger*

**DESCRIPTION** Wing length: 6.5–10mm. A distinctive small hoverfly with pale yellow, curved markings on either side of the oval black abdomen.

**WHERE TO LOOK** Another widespread hoverfly in the three counties whose numbers are boosted by migrants. It can be found in a variety of open habitats, including gardens, visiting flowers.

**WHEN TO LOOK** March–November, peaking twice, in May and July–September.

**SIMILAR SPECIES**

***Scaeva pyrastri*** (see p.27) has similar curved, pale markings on the abdomen, but is a much larger species (wing length: 10.5–12mm).

22

# *Leucozona laternaria*

**DESCRIPTION** Wing length: 7–10mm. A distinctive hoverfly with a pale grey and black abdomen. The thorax has two grey stripes and the scutellum is dark.

**WHERE TO LOOK** A widespread but local hoverfly in the three counties found along or near woodland rides and edges with lush vegetation. It can often be seen visiting umbellifer flowers such as hogweed and angelica in the summer. Try the Lye Valley in Oxford or BBOWT's Dry Sandford Pit nature reserve near Abingdon.

**WHEN TO LOOK** May–September, peaking in July.

**SIMILAR SPECIES**

***Leucozona glaucia*** (wing length: 8–11mm) has blue-grey markings on the abdomen and a yellow scutellum.



# *Leucozona lucorum*

**DESCRIPTION** Wing length: 8–10mm. A bumblebee-like, hairy hoverfly with a cream/pale grey and black abdomen, brown thorax and a yellow scutellum. The wings have a distinctive dark patch.

**WHERE TO LOOK** A widespread hoverfly in the three counties. Look for it along woodland rides and edges, and along hedgerows. It can often be seen visiting flowers such as greater stitchwort, garlic mustard or wild garlic in dappled sunlight in the spring. Try Bernwood Forest in Bucks or these BBOWT woodland sites – Bowdown Woods (Berks), Whitecross Green Wood (Bucks), Dry Sandford Pit and Sydlings Copse (Oxon).

**WHEN TO LOOK** May–August, peaking in May and June.

**SIMILAR SPECIES**

The **Great Pied Hoverfly *Vollucella pellucens***, a similar-looking bumblebee mimic, is much larger (wing length: 10–15.5mm).

# *Melangyna labiatarum*

**DESCRIPTION** Wing length: 6–9mm. A dainty hoverfly with pale yellow spots on the abdomen, a dull brown thorax, pale yellow scutellum and a white face.

**WHERE TO LOOK** A widespread hoverfly of broadleaved woodland in the three counties. Look for it along woodland rides visiting umbellifer flowers such as angelica. Try Bernwood Forest in Bucks or BBOWT's Bowdown Woods nature reserve (Berks).

**WHEN TO LOOK** April–October, peaking twice, in June and August.

**SIMILAR SPECIES**

***Melangyna umbellatorum*** (wing length: 6.5–9mm) looks very similar but has a shiny thorax and very pale, almost white, markings on the abdomen.



# *Meliscaeva cinctella*

**DESCRIPTION** Wing length: 7–10mm. A dainty hoverfly with yellow bands and two yellow spots on the abdomen. The eyes are red-brown.

**WHERE TO LOOK** A common and widespread summer hoverfly in Berks, Bucks and Oxon. Found in various habitats, wherever there are trees or bushes nearby.

**WHEN TO LOOK** April–November, peaking in August and September.

**SIMILAR SPECIES**

***Meliscaeva auricollis*** (wing length: 6–9.5mm) is similar but has triangular markings on the abdomen. It has a longer flight period, February–December.

# *Scaeva pyrastri*

**DESCRIPTION** Wing length: 9–12.5mm. A large black and white hoverfly with distinctive hooked markings on the abdomen.

**WHERE TO LOOK** A common and widespread hoverfly in Berks, Bucks and Oxon. It can be seen visiting flowers on sunny days in various habitats, including scrub and waste ground, meadows and gardens.

**WHEN TO LOOK** May–November, peaking in August.

**SIMILAR SPECIES**

***Scaeva selenitica*** (see below) has yellow hooked markings on the abdomen.



# *Scaeva selenitica*

**DESCRIPTION** Wing length: 10.5–12mm. A large black and yellow hoverfly with distinctive hooked markings on the abdomen.

**WHERE TO LOOK** An uncommon hoverfly in Berks, Bucks and Oxon. Look for it in woodland with conifers, and heathland with conifers nearby. It is often seen at western end of Greenham Common in Berks where it visits flowers along the edge of woodland.

**WHEN TO LOOK** March–November, peaking in June–early September.

**SIMILAR SPECIES**

***Scaeva pyrastri*** (see above) has white hooked markings on the abdomen.

# *Sphaerophoria scripta*

**DESCRIPTION** Wing length: 5–7mm. A small, narrow-bodied hoverfly with yellow markings on the abdomen, a yellow scutellum and yellow markings on the sides of the thorax. The male's (pictured left) abdomen is longer than its wings.
**WHERE TO LOOK** A very common hoverfly in Berks, Bucks and Oxon, the numbers of which are boosted by migrants each year. Look for it in any open, grassy habitat, including flowery waste ground and gardens.
**WHEN TO LOOK** April–November, peaking in July and August.

**SIMILAR SPECIES**

Other ***Sphaerophoria*** species are very similar in size and appearance, but nowhere near as common.



# *Syrphus ribesii* Humming Syrphus

**DESCRIPTION** Wing length: 7–11.5mm. A medium-sized, broad-bodied hoverfly with yellow markings on the abdomen. The female's hind femora are almost entirely yellow, whereas the male's are half yellow. They often settle with their wings constantly vibrating.
**WHERE TO LOOK** A very common and widespread hoverfly in Berks, Bucks and Oxon, found in a variety of habitats including woodlands, hedgerows, waste ground and gardens.
**WHEN TO LOOK** March–November, peaking twice, in late May–early June and July–September.

**SIMILAR SPECIES**

**Glass-winged Syrphus *Syrphus vitripennis*** (wing length: 7–10mm) is very similar but female's hind femur is almost entirely dark.

## *Xanthogramma pedissequum* Superb Ant-hill Hoverfly

**DESCRIPTION** Wing length: 7–10mm. A black and yellow wasp mimic hoverfly with yellow bands and two broad yellow triangles on the abdomen. The wings have a dark patch.

**WHERE TO LOOK** A widespread but local hoverfly in Berks, Bucks and Oxon, found in grasslands and woodland rides. Look for it in the grassland at BBOWT's Dry Sandford Pit nature reserve (Oxon) or along the woodland rides at Bernwood Forest (Bucks).

**WHEN TO LOOK** May–September, peaking in late June and early July.

**SIMILAR SPECIES**

**Barred Ant-hill Hoverfly *Xanthogramma citrofasciatum*** (wing length: 6.5–10mm) has much narrower yellow triangles on the abdomen and the wings have an amber tinge. It prefers well-drained grassland with yellow meadow ants nests.

## *Callicera aurata* Plain Golden-hoverfly

**DESCRIPTION** Wing length: 10–12.5mm. A large hoverfly with a hairy, brassy-coloured abdomen and unmistakeable long antennae with white tips. The larvae live in water-filled cavities in a variety of trees including beech and birch.

**WHERE TO LOOK** A rare hoverfly in our region, with very few records (usually just a single sighting). Most sightings have been from Berkshire, the most recent being from Greenham Common. It can be seen visiting a variety of flowers including hawthorn, bramble, white umbellifers and ivy.

**WHEN TO LOOK** June–September, peaking in July and August.

**SIMILAR SPECIES**

The other two ***Callicera*** species are very rare hoverflies and are not found in our three counties.

## ***Cheilosia illustrata*** Illustrious Cheilosia

**DESCRIPTION** Wing length: 8.5–10mm. A small, hairy black, white and yellow bumblebee mimic hoverfly. Wings clouded.
**WHERE TO LOOK** A widespread midsummer hoverfly in Berks, Bucks and Oxon, found along hedgerows and woodland edges. Look for it on umbellifer flowers at Greenham Common, Bowdown Woods (Berks) and Dry Sandford Pit (Oxon) all BBOWT sites or along the woodland rides at Bernwood Forest in Bucks.
**WHEN TO LOOK** April–September, peaking in July.

**SIMILAR SPECIES**

***Eristalis intricarius*** (see p.47) is another bumblebee mimic of similar size.



## ***Cheilosia pagana*** Pagan Cheilosia

**DESCRIPTION** Wing length: 5–8mm. A small black hoverfly, like many in this genus. The female is, however, instantly recognisable by the large orange antennae. The male is more easily confused with other black *Cheilosia* hoverflies.
**WHERE TO LOOK** A common and widespread hoverfly in the three counties, found along hedgerows, woodlands and open habitats. Look for it in spring on lesser celandine flowers. Later in the year it can be found on umbellifer flowers.
**WHEN TO LOOK** March–October, peaking twice, in May–early June and August.

**SIMILAR SPECIES**

Other black *Cheilosia* hoverflies with orange antennae are the **Marsh-thistle Cheilosia *Cheilosia fraterna*** (wing length: 6–10mm) found on sites with marsh thistle and the **Ragwort Cheilosia *Cheilosia bergenstammi*** (wing length: 7–9mm) found on grassland with plenty of ragwort.

## ***Cheilosia variabilis*** Figwort Cheilosia

**DESCRIPTION** Wing length: 8–10mm. A medium-sized black hoverfly with dark wings and a hairy face.

**WHERE TO LOOK** A common and widespread hoverfly in the three counties, found along woodland rides and clearings. Males are often seen on sunlit perches from where they patrol their territories. Look for it at Bernwood Forest in Bucks and at BBOWT's Chinnor Hill, Dry Sandford Pit and Warburg nature reserves in Oxon.

**WHEN TO LOOK** April–September, peaking in late May–early June.

**SIMILAR SPECIES**

The smaller, all black **Primrose Black-cheilosia *Cheilosia antiqua*** (wing length: 6–8mm) has a shiny face and is found in woodland with primroses.



## ***Cheilosia vulpina***

**DESCRIPTION** Wing length: 7–10mm. A stocky, hairy black hoverfly with a hairy face and yellow markings on dark legs.

**WHERE TO LOOK** A scarce hoverfly in the three counties, found in woodland clearings and on chalk and limestone grasslands with wooded areas nearby. Look for this hoverfly on umbellifer flowers, especially wild parsnip, at BBOWT sites like Dry Sandford Pit and Warburg nature reserves in Oxon.

**WHEN TO LOOK** April–September, peaking twice, in June and August.

**SIMILAR SPECIES**

**Common Thistle-cheilosia *Cheilosia proxima*** (wing length: 6–8.5mm) has white eye hairs is found on marsh thistles in meadows and woodland edges.

# Cheilosia urbana

**DESCRIPTION** Wing length: 6–6.5mm. A small dark brown hoverfly with orange-brown bristles, hairy eyes and orange legs.

**WHERE TO LOOK** A widespread but local hoverfly in Berks, Bucks and Oxon. Found on short-turfed grasslands and heathlands with mouse-ear-hawkweed. In spring, the short grassland on the quarry floor at Dry Sandford Pit nature reserve in Oxon is a good place to look for this small hoverfly.

**WHEN TO LOOK** March–October, peaking in May.

**SIMILAR SPECIES**

**Red-horned Spring-cheilosia *Cheilosia albipila*** (wing length: 8.5–9.5mm) is like a larger version of *Cheilosia urbana* and is found in marshes, wet meadows and fens in the spring.



# Ferdinandea cuprea

**DESCRIPTION** Wing length: 7.5–11mm. A distinctive, medium-sized hoverfly with a metallic brassy abdomen, a black and grey striped thorax and dark spots on the wings.

**WHERE TO LOOK** A widespread woodland hoverfly in Berks, Bucks and Oxon. It is easily overlooked as it usually basks on sunlit trunks of ash or oak trees and rarely visits flowers. Look for this species at BBOWT's nature reserves at Bowdown Woods (Berks) and Lashford Lane Fen (Oxon).

**WHEN TO LOOK** March–November, peaking in June.

**DID YOU KNOW?**

*Ferdinandea cuprea* is especially fond of sap runs from wounds on tree trunks.

## *Rhingia campestris* Common Snout-hoverfly

**DESCRIPTION** Wing length: 6–9.5mm. A medium-sized hoverfly, easily recognised by its long snout, dark striped thorax and red-brown abdomen with a dark marginal line. The larvae are found in cow dung.

**WHERE TO LOOK** A widespread and very common hoverfly in Berks, Bucks and Oxon. It is found in open woodland, woodland edges, hedgerows and can also be found in gardens.

**WHEN TO LOOK** April–October, peaking twice, in late May–early June and late August–September.

**SIMILAR SPECIES**

**Rare Snout-hoverfly *Rhingia rostrata*** (see below) has a more orange abdomen with no obvious dark margin.



## *Rhingia rostrata* Rare Snout-hoverfly

**DESCRIPTION** Wing length: 7.5–9.5mm. A medium-sized hoverfly, easily recognised by its long snout, grey and black striped thorax and bulbous orange abdomen with no obvious marginal line.

**WHERE TO LOOK** A local hoverfly in our three counties. It is found in open woodland and woodland edges. It can be very erratic in appearance, being abundant in some woods one year then disappearing for several years. One of the best places to see this hoverfly is BBOWT's Bowdown Woods nature reserve in Berks.

**WHEN TO LOOK** May–October, peaking twice, in late May–June and September.

**SIMILAR SPECIES**

**Common Snout-hoverfly *Rhingia campestris*** (see above) is much more common and has a darker abdomen with a dark margin.

# *Chrysogaster solstitialis*

**DESCRIPTION** Wing length: 6–7mm. A small, black hoverfly with dark wings and large red eyes.

**WHERE TO LOOK** A common and widespread hoverfly in Berks, Bucks and Oxon. A summer species found in woodland rides and clearings on umbellifer flowers, often several are seen together on each umbel.

**WHEN TO LOOK** Late May–October, peaking in July and August.

**SIMILAR SPECIES**

***Melanogaster hirtella*** (wing length: 5–6mm) has large brown eyes and is found in marshes and water margins.



# *Lejogaster metallina*

**DESCRIPTION** Wing length: 5–6.5mm. A small, metallic green hoverfly with large orange-brown eyes. The female is much broader than the male.

**WHERE TO LOOK** Widespread but local in Berks, Bucks and Oxon, this hoverfly is found in wet meadows, marshes and fens, often visiting buttercups. Look for it near the fen at BBOWT's Dry Sandford Pit nature reserve (Oxon).

**WHEN TO LOOK** May–October, peaking in June–early July.

**SIMILAR SPECIES**

***Orthonevra nobilis*** (see p.43) another small hoverfly (wing length: 4–6mm) found in the fen at Dry Sandford Pit, is copper in colour. It can be mistaken for the larger **Broad Centurian *Chloromyia formosa***, a completely unrelated metallic blue-green or copper-coloured soldierfly.

## *Neoascia podagrica*

**DESCRIPTION** Wing length: 3.5–5mm. An extremely small hoverfly with cloudy wing veins. Both males and females have a bulbous abdomen with a narrow waist. The males (left) have yellowish bands near the base of the abdomen.

**WHERE TO LOOK** Widespread in Berks, Bucks and Oxon, this hoverfly is found in marshes and fens, woodland edges, hedgerows and gardens. It is easily overlooked due to its tiny size and habit of flying low to the ground. Look for it at BBOWT's Dry Sandford Pit where it often sits on wild parsnip umbels and at Parsonage Moor nature reserve (Oxon).

**WHEN TO LOOK** April–November, peaking twice, in late May–early June and August.

**SIMILAR SPECIES**

***Neoascia obliqua*** (wing length: 3.5–5.5mm) also has cloudy wing veins but has oblique angled yellowish bars on the abdomen. They are nowhere near as common and are often seen flying low over butterbur or on the leaves.

## *Orthonevra nobilis*

**DESCRIPTION** Wing length: 4–6mm. A small, metallic copper-coloured hoverfly with black legs.

**WHERE TO LOOK** A local hoverfly in Berks, Bucks and Oxon, it is found in fens, marshes and damp woodland clearings, but never in any great number. Like the previous species it also keeps low to the ground. Look for it at BBOWT's Dry Sandford Pit and Parsonage Moor nature reserves (Oxon).

**WHEN TO LOOK** May–September, peaking in late June–early August.

**SIMILAR SPECIES**

***Lejogaster metallica*** (see p.41) is slightly larger (wing length: 5–6.5mm) and metallic green-bronze and can be found in similar habitats.

44

# *Eristalinus sepulchralis*

**DESCRIPTION** Wing length: 6.5–8mm. A dumpy black hoverfly with distinctive hairy, mottled eyes, a dark-grey striped thorax and a glossy, tapered abdomen.

**WHERE TO LOOK** A widespread but local hoverfly in Berks, Bucks and Oxon, it is found in fens, marshes and near ponds. Look for it at BBOWT's Dry Sandford Pit and Parsonage Moor nature reserves (Oxon).

**WHEN TO LOOK** April–October, peaking in July.

**SIMILAR SPECIES**

***Eristalinus aeneus*** (wing length: 6–9.5mm) is the only other species with mottled eyes, but it does not occur in our area as it is a coastal species.



# *Eristalis arbustorum*

**DESCRIPTION** Wing length: 7–10mm. The genus *Eristalis* are honeybee mimics and includes some of our commonest hoverflies. This species is one of the smaller members of the genus and has a yellowish face. It is typically dark with a large yellow patch on either side of the abdomen (left), but can have an almost entirely dark abdomen (below).

**WHERE TO LOOK** A widespread hoverfly in Berks, Bucks and Oxon, it is found in wetland habitats and in gardens, and can often be seen visiting flowers.

**WHEN TO LOOK** April–November, peaking in July and August.

**SIMILAR SPECIES**

***Eristalis horticola*** (see p.47) is another widespread member of the genus. Slightly larger (wing length: 8–11.5mm) it is found in a wide range of habitats.

# *Eristalis horticola*

**DESCRIPTION** Wing length: 8–11.5mm. A particularly brightly coloured orange and dark brown *Eristalis* hoverfly. It has noticeable dark marks across the centre of the wings.

**WHERE TO LOOK** A widespread hoverfly in Berks, Bucks and Oxon, though not particularly common. It is found in a variety of habitats and always near trees and bushes. It is often seen visiting flowers

**WHEN TO LOOK** April–October, peaking in July.

**SIMILAR SPECIES**

***Eristalis arbustorum*** (see p.45) is a smaller hoverfly (wing length: 7–10mm) and does not have the dark wing marks.



# *Eristalis intricarius*

**DESCRIPTION** Wing length: 8–12mm. A bumblebee mimic, this is the hairiest *Eristalis* of our hoverflies. The male (below) is dark with a orange-brown tail and the female (left) is larger and has a white tail.

**WHERE TO LOOK** A widespread hoverfly in the three counties, though never seen in any great number. It can be found in marshes, fens, wet heaths and woodlands, especially where willows are present. In spring they can be seen visiting blackthorn blossom and willow catkins, and in the summer can be seen on a variety of flowers. Greenham Common in Berks and Dry Sandford Pit nature reserve in Oxon are two places where you might see them.

**WHEN TO LOOK** March–October, peaking in July.

**SIMILAR SPECIES**

Two bumblebee mimics the **Illustrious Cheilosia *Cheilosia illustrata*** (see p.33), smaller, (wing length: 8.5–10mm) and the larger ***Volucella bombylans*** (wing length: 8–14mm) are both found in our area.

## *Eristalis pertinax*

**DESCRIPTION** Wing length: 8–13mm. A honeybee mimic and one of the largest members of the *Eristalis* genus. It has a distinctive dark, tapered abdomen which it constantly twitches in a similar fashion to that of a honeybee.

**WHERE TO LOOK** A very common and widespread hoverfly in Berks, Bucks and Oxon. It can be one of the most abundant hoverflies seen in woodlands wherever there are suitable flowers such as hawthorn blossom. The males are often seen along woodland rides hovering and fending off rivals in their small territories. They can sometimes be seen in gardens with woodland nearby.

**WHEN TO LOOK** April–November, peaking in May and August.

**SIMILAR SPECIES**

The Drone Fly ***Eristalis tenax*** (see below) is of a similar size but does not have the tapered abdomen and is slightly hairier and more variably patterned.

## *Eristalis tenax* Drone Fly

**DESCRIPTION** Wing length: 10–13mm. The Drone Fly gets it common name because of its resemblance to a drone honeybee, and like a honeybee, even twitches its abdomen in a similar fashion.

**WHERE TO LOOK** A very common and widespread hoverfly of woodland, hedgerows, scrub and gardens in the three counties. One of our most abundant hoverflies in high summer, especially in gardens where it can be seen visiting flowers such as Michaelmas daisies.

**WHEN TO LOOK** All year, peaking in August and September. The females hibernate in crevices and buildings.

**SIMILAR SPECIES**

***Eristalis pertinax*** (see above) is of a similar size, but has dark, narrowly tapered abdomen.

# *Helophilus pendulus* Common Stripe-back

**DESCRIPTION** Wing length: 8.5–11mm. This hoverfly is said to be our commonest species. The thorax has pale yellow to cream stripes and the abdomen is attractively marked with large yellow patches.

**WHERE TO LOOK** A very common and widespread hoverfly in Berks, Bucks and Oxon. They can be found in various habitats, including gardens, particularly with ditches, shallow ponds and pools nearby, and can even be seen in numbers around muddy puddles.

**WHEN TO LOOK** April–November, peaking in July.

**SIMILAR SPECIES**

**Marsh Stripe-back *Helophilus hybridus*** (wing length: 8.5–11mm) is similar but duller in colour and has the two yellow patches on each side of the abdomen merged together. ***Parhelophilus frutetorum*** (see p.53 is similarly patterned but more orange and much smaller (wing length: 7–9mm)

# *Helophilus trivittatus* Large Stripe-back

**DESCRIPTION** Wing length: 10–12mm. A large, attractively marked hoverfly. It has a yellow face, striped thorax and large, bright yellow patches on the abdomen, though these sometimes can be darker (see below). The males have a pale 'W'-shaped marking at the tip of the abdomen.

**WHERE TO LOOK** A widespread hoverfly in Berks, Bucks and Oxon, though not common. It is found in various habitats including marshes and fens. It is often seen on umbellifers, thistles and hemp-agrimony. Look for it at BBOWT's Dry Sandford Pit and Parsonage Moor nature reserves.

**WHEN TO LOOK** May–October, peaking in August.

**SIMILAR SPECIES**

The **Common Stripe-back *Helophilus pendulus*** (see above) is slightly smaller (wing length: 8.5–11mm) and has duller yellow patches on the abdomen.

# *Myathropa florea*

**DESCRIPTION** Wing length: 7–12mm. A medium-sized, black and yellow wasp mimic hoverfly. The dark thorax is distinctive with a soft greyish patterning and patches of yellow hairs.

**WHERE TO LOOK** A common and widespread hoverfly of wooded areas in the three counties. It is particularly fond of umbellifer and ivy flowers. Males can often be seen hovering in beams of sunlight. Look for it on any BBOWT woodland nature reserve or along hedgerows.

**WHEN TO LOOK** May–October, peaking in July and August.

**SIMILAR SPECIES**

This hoverfly could possibly be confused with poorly marked ***Eristalis*** species (see pp.45–49), but the distinctively marked thorax should be enough to confirm identification.

# *Parhelophilus frutetorum*

**DESCRIPTION** Wing length: 7–9mm. A small, bright orange and black hoverfly. The female is furrier and more orange than the male.

**WHERE TO LOOK** A widespread hoverfly in Berks, Bucks and Oxon. Although this hoverfly is a wetland species found along pool sides and ditches, it is equally likely to be seen around scrubby woodland edges and by patches of tall vegetation. The woodland edge at BBOWT's Bowdown Woods is a favourite spot for this bright little hoverfly as is the adjacent Greenham Common (both in Berks) which has plenty of small pools and wet areas. Ewelme Watercress Beds in Oxon is also worth checking out.

**WHEN TO LOOK** April–September, peaking in June.

**SIMILAR SPECIES**

The less common ***Parhelophilus versicolor*** (wing length: 7–9mm) is less hairy, but is rarely seen away from its wetland habitats, unlike *Parhelophilus frutetorum*.

# ***Merodon equestris*** Large Narcissus Fly

**DESCRIPTION** Wing length: 8.5–10mm. A bumblebee mimic hoverfly that has a variety of colour forms – form *equestris* (left), form *validus* (only females) (below right), form *narcissi* (see box) and form *transversalis* (see p.2). The larvae attack narcissus (hence the common name), bluebell and daffodil bulbs.

**WHERE TO LOOK** A common and widespread hoverfly in Berks, Bucks and Oxon. It is found in various habitats including bluebell woods, meadows and gardens, all with plentiful flowers. It is especially fond of dandelions.

**WHEN TO LOOK** April–September, peaking in late May and June.

**Note:** form *narcissi*

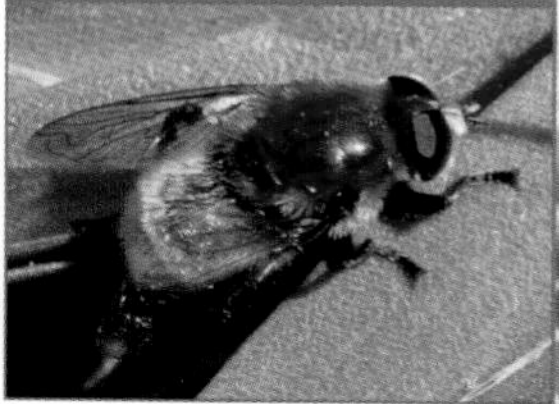



# ***Pipizella viduata***

**DESCRIPTION** Wing length: 4–5mm. A small black hoverfly with long antennae that mimics small dark *Lasioglossum* solitary bees.

**WHERE TO LOOK** A widespread hoverfly in Berks, Bucks and Oxon, but easily overlooked owing to its small size and dark coloration. It is found in dry chalk and limestone grasslands with bedstraws and wild parsnip, heathland, woodland rides and disturbed ground. BBOWT's Dry Sandford Pit and Warburg nature reserves (Oxon), both with plentiful wild parsnip, are worth checking out for this small hoverfly.

**WHEN TO LOOK** May–September, peaking in late June and early July.

**SIMILAR SPECIES**

The much scarcer ***Pipizella virens*** (wing length: 5–6mm) has slightly longer antennae and more pointed wings. It is found in the same range of habitats.

56

# Pipiza noctiluca

**DESCRIPTION** Wing length: 6.5–8mm. A medium-sized black hoverfly. The female (left) has two dull yellowish spots on the abdomen and dark clouded markings on the wings. The male (below) has black hairs on the face, no spots on the abdomen and no cloudy markings on the wings.

**WHERE TO LOOK** A widespread hoverfly in Berks, Bucks and Oxon. It is found along woodland edges and hedgerows, often sunbathing on leaves or visiting flowers of brambles and umbellifers. Look for it at Bernwood Forest in Bucks or at BBOWT's Bowdown Woods nature reserve.

**WHEN TO LOOK** April–September, peaking in late May and early June.

**SIMILAR SPECIES**

***Pipiza austriaca*** (see below) is of a similar size but has no spots on the abdomen and has well-defined, dark brown patches on the wings in both the male and the female.



# Pipiza austriaca

**DESCRIPTION** Wing length: 6–8mm. A distinctive, medium-sized, black hoverfly with bands of whitish hairs. Both male and female have well-defined dark brown, cloudy patches on the wings.

**WHERE TO LOOK** A widespread but local hoverfly in Berks, Bucks and Oxon. It can be found along woodland edges and rides often visiting umbellifer flowers, and in meadows where it visits buttercups and yellow rattle. Look for it at BBOWT's Whitecross Green Wood (Bucks) and Bowdown Woods (Berks). It can also be seen at Bernwood Forest in Bucks.

**WHEN TO LOOK** May–September, peaking in June.

**SIMILAR SPECIES**

Only the female ***Pipiza noctiluca*** (see above) has dark wing patches, she also has two yellowish spots on the abdomen.

58

# *Sericomyia silentis*

**DESCRIPTION** Wing length: 9.5–14mm. A large, conspicuous wasp mimic with wedge-shaped yellow bars on a broad abdomen.

**WHERE TO LOOK** A scarce and local hoverfly in Berks, Bucks and Oxon. Found mainly on wet heathland and bogs, it can also occur along woodland rides, meadows and sometimes gardens where the underlying soils are acid. It will settle briefly on foliage and on flowers such as devil's-bit scabious. It is most likely to be seen in Berkshire at sites like Greenham Common, Snelsmore Common, BBOWT's Bowdown Woods, Decoy Heath and Wildmoor Heath nature reserves.

**WHEN TO LOOK** May–November, peaking in July.

**SIMILAR SPECIES**

***Sericomyia lappona*** (wing length: 9–11mm) is smaller with narrower bands on the abdomen. Also found on heathlands but is much rarer.



# *Volucella bombylans*

**DESCRIPTION** Wing length: 8–14mm. A large, conspicuous hairy bumblebee mimic with feathery antennae and dark patches on the wings. This hoverfly has two main colour forms – one is black with an orange-red tip to the abdomen (mimics the red-tailed bumblebee) and the other is black with a yellow band on the thorax and a white tip to the abdomen (mimics white-tailed bumblebee). The larvae develop in bumblebee nests.

**WHERE TO LOOK** A common and widespread hoverfly in Berks, Bucks and Oxon. Found along woodland edges and scrubby grassland. Any number of woodland sites in the three counties will have this hoverfly, so check out any bumblebee you see carefully, it might well turn out to be a hoverfly instead.

**WHEN TO LOOK** May–September, peaking in June.

**SIMILAR SPECIES**

***Eristalis intricarius*** (see p.47) is a smaller bumblebee mimic (wing length: 8–12mm).

# *Volucella inflata*

**DESCRIPTION** Wing length: 11–13mm. A large, conspicuous hoverfly with a very rounded black and orange abdomen and dark patches on the wings.

**WHERE TO LOOK** A very scarce hoverfly in Berks, Bucks and Oxon. It is found in ancient woodlands and is associated with sap runs on trees where females have been seen egglaying. BBOWT's Bowdown Woods nature reserve (Berks) is worth checking out in the summer where you might catch a glimpse of the elusive hoverfly on a dog-rose bloom or sunning itself on a sunlit tree trunk.

**WHEN TO LOOK** May–September, peaking in late June–July.

**SIMILAR SPECIES**

**Great Pied Hoverfly *Volucella pellucens*** (see below) is very similar in shape but has a black and white abdomen.



# *Volucella pellucens* Great Pied Hoverfly

**DESCRIPTION** Wing length: 10–15.5mm. A large, conspicuous hoverfly with a very rounded black and white abdomen and dark patches on the wings. The larvae live in the nests of wasps.

**WHERE TO LOOK** A widespread hoverfly in Berks, Bucks and Oxon. It is found in woodland glades and rides and in copses. They are commonly seen visiting bramble flowers but can also be seen on umbellifer flowers as well. You will find this hoverfly along the rides in Bernwood Forest (Bucks) and at BBOWT's nature reserves at Bowdown Woods (Berks) and Dry Sandford Pit (Oxon), but many other woodland sites will have this conspicuous species as well.

**WHEN TO LOOK** May–September, peaking in late June–July.

**SIMILAR SPECIES**

***Leucozona lucorum*** (see p.23) seen in spring along woodland rides, has a large white patch on the abdomen, but is much smaller (wing length: 8–10mm).

62

# *Volucella inanis*

**DESCRIPTION** Wing length: 12–14mm. A large, conspicuous wasp mimic hoverfly with yellow-orange bands on the abdomen and dark patches on the wings. The larvae live in the nests of the German wasp and the hornet.

**WHERE TO LOOK** A widespread but local hoverfly in Berks, Bucks and Oxon. It is found along woodland edges, waste ground and gardens. They visit thistles and ivy flowers in particular. Good sites to see this large hornet-like hoverfly are Greenham Common and Bowdown Woods nature reserve (Berks), Dry Sandford Pit nature reserve (Oxon), all BBOWT sites, and Bernwood Forest in Bucks.

**WHEN TO LOOK** June–September, peaking in early August.

**SIMILAR SPECIES**

**Hornet Hoverfly *Volucella zonaria*** (see below) is much larger (wing length: 15.5–19.5mm) and is occasionally glimpsed in our area.

# *Volucella zonaria* Hornet Hoverfly

**DESCRIPTION** Wing length: 15.5–19.5mm. Britain's largest hoverfly and hornet mimic, this hoverfly arrived in Britain on the south coast in the 1930s and soon became established in the London area. It has a brown thorax and a black and yellow abdomen, and at first glance does indeed look like a hornet. The larvae live in wasps nests.

**WHERE TO LOOK** A widespread hoverfly in Berks, Bucks and Oxon, having spread north and west from the London area. It can be seen along woodland edges but is equally at home on waste ground and in gardens where it is often seen on buddleia.

**WHEN TO LOOK** May–November, peaking in August.

**SIMILAR SPECIES**

Only really likely to be confused with ***Volucella inanis*** (see above) which is smaller (wing length: 12–14mm).

# *Brachypalpoides lentus*

**DESCRIPTION** Wing length: 10–12mm. A distinctive, easily identified, sawfly mimic hoverfly. Mainly black with a large area of red on the abdomen.

**WHERE TO LOOK** An uncommon hoverfly in Berks, Bucks and Oxon. Found in woodlands, especially with beech, where it is often found in shady spots near fallen trees or in hollows of beech and oak trees. Look out for it in Windsor Forest or at BBOWT's Bowdown Woods nature reserve in Berks.

**WHEN TO LOOK** April–August, peaking in June.

**SIMILAR SPECIES**

No other hoverfly has the striking blood-red abdomen, although ***Xylota segnis*** can look similar but is smaller (wing length: 7–9.5mm) and has a black-tipped yellow abdomen.



# *Calyprobola speciosa*

**DESCRIPTION** Wing length: 11–12.5mm. A handsome and distinctive wasp mimic hoverfly with a metallic greenish-brown abdomen, orange legs and amber wings. The sound of its buzzing flight is reminiscent of a hornet.

**WHERE TO LOOK** A very rare hoverfly in our area (and in the country) where it is only found in Windsor Forest and Great Park where it is seen on the ground near or on ancient beech trees and more rarely oak. The only other area in Britain for this spectacular species is the New Forest in Hampshire.

**WHEN TO LOOK** May–July, peaking in June.

**DID YOU KNOW?**

This very rare hoverfly is a Red List RDB1 species and is endangered in Britain (found in five or fewer 10-km squares and in danger of extinction).

# *Criorhina floccosa*

**DESCRIPTION** Wing length: 10–13mm. A bumblebee mimic hoverfly resembling a common carder bumblebee. It has an orange furry thorax, a white face and very long legs.

**WHERE TO LOOK** A local hoverfly in Berks, Bucks and Oxon, found in woodlands and along hedgerows. It can be seen flying low to the ground around the bases of trees, on which they settle before flying off again only to return to the same spot some moments later. Bernwood Forest in Bucks or Windsor Forest in Berks are worth a visit for this species.

**WHEN TO LOOK** April–August, peaking in late May–early June.

**SIMILAR SPECIES**

The smaller ***Chriorhina berberina*** (wing length: 9.5–11.5mm) is similar in colour and is the commonest species in the genus. It is found in ancient woodlands.



# *Criorhina ranunculi*

**DESCRIPTION** Wing length: 11–14mm. A black-haired bumblebee mimic hoverfly, with a white tail or a reddish-orange tail.

**WHERE TO LOOK** A scarce hoverfly in Berks, Bucks and Oxon. A spring hoverfly of woodlands and wooded heathland, where it can be seen visiting the flowers of blackthorn, wild cherry or willows, or flying low around the trunks of trees. Good sites to see this nationally scarce hoverfly are Windsor Forest and BBOWT's Bowdown Woods nature reserve in Berks or the woodlands around BBOWT's Parsonage Moor nature reserve in Oxon.

**WHEN TO LOOK** March–July, peaking in late April.

**SIMILAR SPECIES**

The red-tailed form of ***Volucalla bombylans*** is of similar size but has distinctive black markings on the wings and usually flies later in the year.

# *Syritta pipiens*

**DESCRIPTION** Wing length: 5–7mm. A small hoverfly with a down-curved, orange or grey spotted abdomen. The hind legs have noticeably swollen femora.

**WHERE TO LOOK** A common and widespread hoverfly in Berks, Bucks and Oxon. It is found in various habitats including gardens, heathland, hedgerows, meadows and marshes where it can often be seen in abundance visiting a range of flowers throughout spring and summer and on ivy flowers later in the year.

**WHEN TO LOOK** April–November, peaking in July and August.

**SIMILAR SPECIES**

The slightly larger ***Tropida scita*** (see below) also has swollen hind femora, but has an orange abdomen and is found in wetlands with reedbeds.



# *Tropida scita*

**DESCRIPTION** Wing length: 5.5–8mm. A small hoverfly with an orange abdomen and with swollen femora with a sharp tooth on the hind legs.

**WHERE TO LOOK** A widespread hoverfly in Berks, Bucks and Oxon, where it can be found in marshes and fens with reedbeds nearby. It has a slow, silent flight and will visit wetland flowers. Look for it wherever there are reedbeds nearby. Good sites to see this hoverfly are the BBOWT's Cothill fens natures reserves (Dry Sandford Pit, Lashford Lane Fen and Parsonage Moor) in Oxon and Greenham Common and Thatcham Reedbeds in Berks.

**WHEN TO LOOK** May–October, peaking in June and July.

**SIMILAR SPECIES**

***Xylota segnis*** (see p.71) is larger (wing length: 7–9.5mm), similar in shape but lacks the sharp tooth on the hind femora.

# *Xylota segnis*

**DESCRIPTION** Wing length: 7–9.5mm. A medium-sized hoverfly with an orange-yellow abdomen with a black tip. It has long legs with swollen hind femora.

**WHERE TO LOOK** A common and widespread hoverfly in the three counties. Found in mixed woodland, hedgerows and gardens where it can be seen sunbathing on foliage, logs or stone. If disturbed it quickly darts off but usually returns to the same spot a few moments later. There are many woodland sites where you can see this sawfly-like hoverfly.

**WHEN TO LOOK** April–November, peaking in June and July.

**SIMILAR SPECIES**

***Xylota sylvarum*** (see below) is larger (wing length: 7–12mm) and has a black abdomen with a yellow tip.



# *Xylota sylvarum*

**DESCRIPTION** Wing length: 7–12mm. A large sawfly mimic hoverfly with a black abdomen with a golden yellow tip. The hind legs have swollen femora.

**WHERE TO LOOK** A widespread hoverfly in Berks, Bucks and Oxon, though not so common as the previous species. Look for it in woodlands or in areas with woodland close by. They are frequently seen sunbathing on foliage and logs or quickly walking over them. Like the previous species it is nervous and will dart off if disturbed, but will soon return to its favourite sunbathing spot a few moments later.

**WHEN TO LOOK** May–October, peaking in late June and July.

**SIMILAR SPECIES**

The more common ***Xylota segnis*** (see above) is smaller (wing length: 7–9.5mm) and has a yellow abdomen with a black tip.

# Species index